discovermore

Transportation

Focus on Trains

Doug Bradley

I0869082

Britannica

Educational Publishing

IN ASSOCIATION WITH

THE ROSEN PUBLISHING GROUP

Published in 2026 by Britannica Educational Publishing (a trademark of Encyclopædia Britannica, Inc.)
in association with The Rosen Publishing Group, Inc.
2544 Clinton Street, Buffalo, NY 14224

Distributed exclusively by Rosen Publishing.
To see additional Britannica Educational Publishing titles, go to rosenpublishing.com.

Editor: Greg Roza
Book Design: Michael Flynn

Photo Credits: Cover Denis Belitsky/Shutterstock.com; (series background) Dai Yim/Shutterstock.com; p. 4 Guitar photographer/Shutterstock.com; p. 5 Christopher O'Donnell/Shutterstock.com; p. 6 nexus 7/Shutterstock.com; p. 7 DedMityay/Shutterstock.com; p. 8 Aleksandar Malivuk/Shutterstock.com; p. 9 PHOTOGRAPHY IS ON/Shutterstock.com; p. 11 (top) https://commons.wikimedia.org/wiki/File:First_Horsecar_in_Manchester,_NH.jpg; p. 11 (bottom) Alex Segre/Shutterstock.com; p. 12 https://commons.wikimedia.org/wiki/File:The_Last_Spike_1869.jpg; p. 13 aaddyy/Shutterstock.com; p. 14 People Image Studio/Shutterstock.com; p. 15 Nicola Pulham/Shutterstock.com; p. 16 Altrendo Image/Shutterstock.com; p. 17 Erman Gunes/Shutterstock.com; p. 18 EL BANCO04/Shutterstock.com; p. 19 CIS/Shutterstock.com; p. 20 Moyashi Kanagawa/Shutterstock.com; p. 21 06photo/Shutterstock.com; p. 23 (top) Mark Agnor/Shutterstock.com; p. 23 (bottom) Arthur E Gurmankin/Shutterstock.com; p. 25 (top) dvande/Shutterstock.com; p. 25 (bottom) Krysja/Shutterstock.com; p. 26 R.M. Nunes/Shutterstock.com; p. 27 Jiaye Liu/Shutterstock.com; p. 28 Tom Eversley/Shutterstock.com; p. 29 Chubykin Arkady/Shutterstock.com.

Library of Congress Cataloging-in-Publication Data

Names: Bradley, Doug, 1971- author
Title: Focus on trains / Doug Bradley.
Description: [Buffalo, New York] : Britannica Educational Publishing, [2026] | Series: Discover more: transportation | Includes index.
Identifiers: LCCN 2025037256 (print) | LCCN 2025037257 (ebook) | ISBN 9781641905572 library binding | ISBN 9781641905565 paperback | ISBN 9781641905589 ebook
Subjects: LCSH: Railroad trains--Juvenile literature | Railroad trains--History--Juvenile literature | Railroad trains--Technological innovations--Juvenile literature | Railroads--Juvenile literature | Railroads--History--Juvenile literature | Railroads--Technological innovations--Juvenile literature | CYAC: Railroad trains | Railroads
Classification: LCC TF148 .B694 2026 (print) | LCC TF148 (ebook)
LC record available at https://lccn.loc.gov/2025037256
LC ebook record available at https://lccn.loc.gov/2025037257

Manufactured in the United States of America

Some of the images in this book illustrate individuals who are models. The depictions do not imply actual situations or events.

CPSIA Compliance Information: Batch #CWBRIT26. For further information contact Rosen Publishing at 1-800-237-9932.

Contents

All Aboard!

Have you ever taken a trip by train? It's one of the fastest ways for many people to travel together over land. Trains travel along a path of two metal rails, or tracks. Each section of the train is linked together. The wheels on the bottom have a slightly curved outer edge. That is what makes them stay on the rails.

A trip by train can be both relaxing and exciting. Look at that view!

Freight trains are always on the move all over the country.

There are two types of trains: passenger and freight. Passenger trains carry people. Freight trains carry products. Railroads have been essential to the spread of people settling across many countries, including the United States. Today, trains help people travel across almost every country in the world. Railroads serve many thousands of communities, from big cities to tiny villages in faraway areas.

Consider This

Why would one travel by train rather than by car?
Why would one choose a train over an airplane?

On the Tracks

Railroads are the most **efficient** method of land transportation. They require the least amount of fuel and work by people. Railroads are designed to move large amounts of passengers or freight over long distances.

A locomotive is the vehicle that pulls most trains. It can be powered by an engine that burns diesel fuel or by electricity. The electricity comes from a third rail or an overhead wire.

Diesel fuel is similar to gasoline. However, it produces more energy than gas, and is often more expensive.

Electrically-powered trains are more common in populated areas.

The train operator, called the driver or engineer, usually sits in the locomotive. Some modern trains have computers that help guide the train and control its speed. The engineer uses the computers and other devices to run the train. Those devices include a brake, throttle, and reversing gear. The brake controls slowing and stopping. The throttle controls the locomotive's speed. The reversing gear allows the train to back up.

WORD WISE

SOMETHING THAT WORKS WELL AND DOES NOT WASTE TIME OR RESOURCES IS EFFICIENT.

The locomotive pulls connected sections commonly called cars. The underside of a train car is called an undercarriage. It holds the wheels. Under each end of a car are couplers. They look like hooks. Their job is to connect the cars.

A train runs on rails, or tracks. They guide the train and help spread the weight of the train on the ground. Tracks usually have two parallel steel rails. They are spaced 4 feet 8.5 inches (1.44 meters) apart. The rails are connected by wood or concrete pieces called crossties. The rails are bolted down to the crossties. The ties sit on loose gravel or another material that provides a cushioned, steady foundation.

A worker examines the wheels of a train car to make sure they are working properly.

Railroad switches allows trains traveling to different places to use the same tracks. They're kind of like road intersections.

The rails have switches at some points. This is usually where two or more tracks come together. Switches allow trains to take different paths as they move forward.

Consider This

Old trains were controlled by human workers. Today, many trains are controlled by computers, although they still need an engineer or driver. Why is it important to have people as well as computers controlling trains?

The History of Trains

The first trains were just carts pulled by horses along tracks. Europeans began using this early type of animal-led railroad in the 1500s. It allowed them to carry heavy loads to and from mines, where workers dug up coal and other minerals.

In 1803 British engineer Richard Trevithick planned and built a locomotive that ran on steam power. Mining companies were the first to use it. In the 1820s another British inventor, George Stephenson, made the first steam train to carry products and passengers. It was called the Rocket. Soon railroads were being built all over Europe. George Stephenson and his son Robert continued to improve the steam locomotive design. Stephenson also helped design and build the Liverpool and Manchester Railway in England.

Shown here is the first horse-drawn train to set up service in Manchester, New Hampshire, in 1877.

compareandcontrast

Steam engines replaced horse-drawn locomotives. Why do you think this is? Do you think there were any benefits to a horse-drawn locomotive compared to one powered by a steam engine?

Stephenson's Rocket steam locomotive is on display at the Science Museum in London.

The first railroads in the United States connected cities in the eastern part of the country. In 1862 the government gave money to two companies to build a route across the country. The Central Pacific laid track eastward from Sacramento, California. The Union Pacific built westward from Omaha, Nebraska. They met in Promontory, Utah, in May 1869. This was the first transcontinental railroad.

This painting from 1881, titled *The Last Spike*, shows the ceremony held as the last spike was driven to complete the first transcontinental railroad in Promontory, Utah.

Electric locomotives such as the one shown here are usually better for the environment than diesel locomotives.

Electric locomotives were introduced in 1895. They receive power through overhead wires or third rails connected to central power stations. In 1925 U.S. railroads started to use an electric diesel locomotive that carried its own power source.

Diesel engines are still used all over the world. However, some of the busiest railroads—including those in the United States, Europe, Russia, and Japan—use electricity from a central source. Even steam locomotives are still used in parts of the world.

Consider This

As railroads stretched across North America, many Native American groups were forced to move from their homelands. Do you think this was fair?

Branching Out

The first U.S. railroads started near the East Coast. That's where most of the county's ports were located and where many goods traveled in and out of the country. Trains connected these cities, so more people could receive these goods. By the mid-1850s, railroad tracks had been extended to reach other cities to the west. Some tracks even connected parts of the United States to Canada.

Ports are busy places! They are hubs where ships, trains, trucks, and planes converge, or meet.

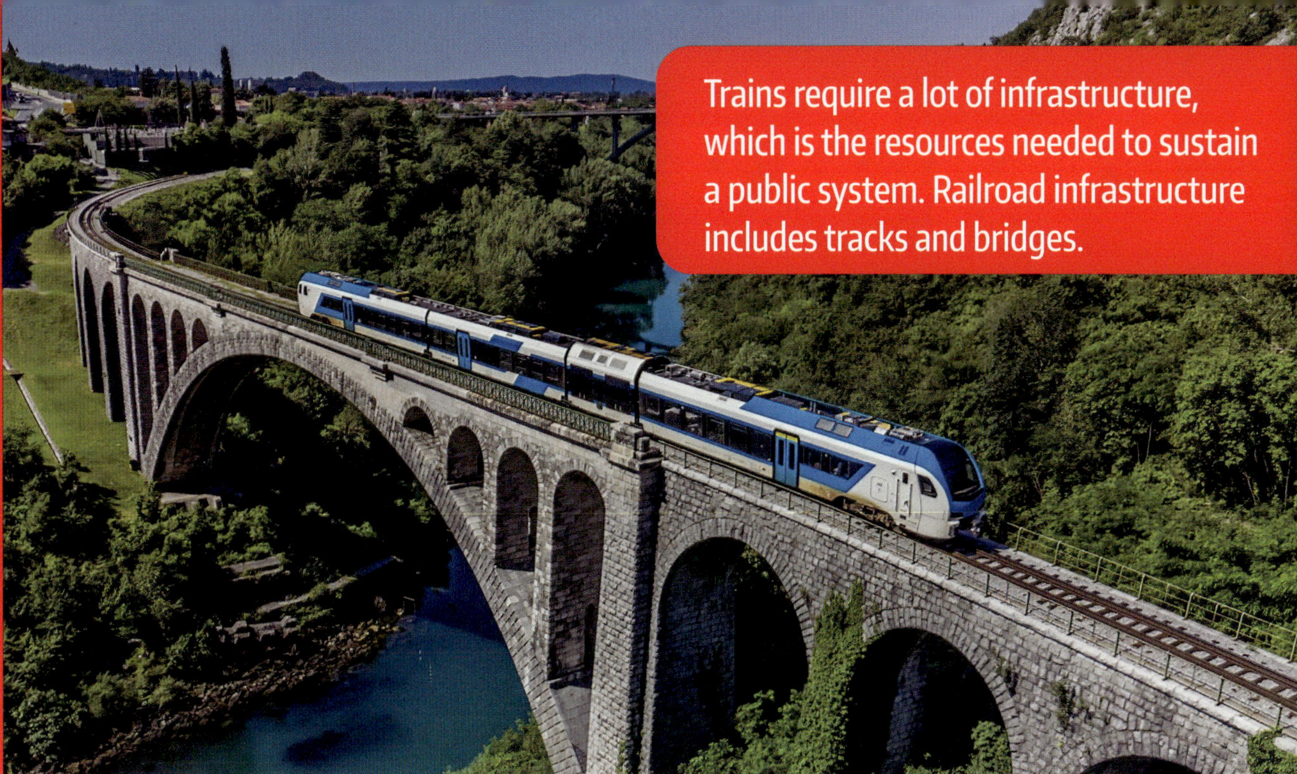

Trains require a lot of infrastructure, which is the resources needed to sustain a public system. Railroad infrastructure includes tracks and bridges.

Building railroads takes a lot of work. First someone has to decide what route the railroad should take. There are many other issues to address as well. Trees may have to be cut down. Tunnels need to be cut through mountains. Bridges need to be built to span rivers. When the land is ready, the track is laid by hand or by machines. Railroad tracks require constant observation and frequent repairs.

Consider This

The United States has far fewer railroads than other countries, especially those in Europe and Asia. Why might that be?

People on the Move

The cars forming a passenger train are filled with seats. Trains that cover long distances have dining cars, baggage cars, and even sleeping cars with beds. The speed of many large passenger trains is about 100 miles (160 km) per hour. However, some trains in Europe and Japan can go much faster.

Depending on the railroad and the length of the trip, train travelers may enjoy a meal in a fancy dining car!

Subways are underground trains that carry passengers around large, crowded cities.

Other trains carry people shorter distances. These trains often do not have a separate locomotive. Instead, some or all of the cars have electric motors under the floor. The train operator sits in the front car. These include commuter trains (trains that carry workers into a city), subways (underground trains), and monorails (trains that run on one rail).

Commuter trains have cars that carry as many people as possible. Some commuter cars have two levels of seats and can carry as many as 150 passengers.

Consider This

Why do you think many cities create underground railroads, or subways?

Bullet Trains

Train technology has kept rolling on! In the 1900s some countries began building trains that could go much faster than ordinary trains. Japan built the world's first high-speed train. It began operating on October 1, 1964. The name of the train is the Shinkansen. It was built to travel from Tokyo to Osaka but has since been extended. These trains can travel up to 177 miles (285 km) per hour.

The Shinkansen speeds past Mount Fuji on the island of Honshu, Japan.

TGV is short for *train à grande vitesse*, which is French for "high-speed train."

France's high-speed trains are known as the TGV. One train set a new world speed record of 320 miles (515 km) per hour in 1990, but the trains usually run slower than that. The TGV trains are powered by front and rear electric locomotives.

Many other countries have high-speed trains or are planning to build them. In the United States, the Acela is a high-speed train service that runs from Boston, Massachusetts, to Washington, D.C. These trains travel up to 150 miles (241 km) per hour. Many states, including new York, Florida, California, and Texas, have plans for future high-speed rail lines.

Consider This

The Shinkansen and high-speed trains similar to it are nicknamed "bullet trains." Can you tell why?

Maglev Trains

The fastest trains in the world are magnetic levitation, or maglev, trains. Maglevs do not have wheels moving on rails. Instead, magnetic coils in the surface of the track create a system that lifts the vehicles and moves them forward. The ride is smoother than an ordinary train. Speeding up and slowing down are faster because the train's bottom never touches the track. Bad weather usually has no effect on this train's movement.

The Shanghai maglev train travels a total distance of about 18.6 miles (30 km) in just over 8 minutes.

Maglev tracks are different from traditional railroad tracks. This means a new infrastructure needs to be created for them.

In 2004 China's Shanghai Transrapid became the world's first commercial maglev train line—and the world's fastest at the time. It operates in China's most populated city, Shanghai. It runs at a speed of 267 miles (430 km) per hour.

Japan has been developing maglev trains, called L0 trains, that have reached a top speed of 375 miles (603 km) per hour. When finished, this train will take passengers from Tokyo to Osaka in just over an hour.

Consider This

Several scientists had experimented with maglev technology as early as 1900. Why do you think it so long to become popular?

Moving Right Along

Freight trains carry goods rather than people. Boxcars carry packaged goods or other freight that must be protected from the weather. Some are cushioned to protect fragile shipments from damage. Open-top hopper cars can carry minerals, such as coal or iron ore. Covered hopper cars carry grain. Tank cars carry liquids. Car carriers carry automobiles. Flatcars are open with no side panels. They can carry large, heavy materials, such as lumber and construction equipment.

Companies that run freight trains compete with highway trucking companies to get other companies to pay them to move their products. Some freight trains now have cars stacked on top of each other and specialized cars to carry liquid chemicals or refrigerated items.

Freight trains can have 150 cars or more. These trains usually move more slowly than passenger trains because they pull so much weight.

Hundreds of hopper cars loaded with coal leave the coal mine.

compare and contrast

Which do you think is safer to haul: single cars, or double-stacked cars? Which do you think brings in more money for shipping companies?

A double-stack container train can carry double the amount of goods.

Tracking Trains

The world has more than 807,800 miles (1.3 million km) of railroad tracks. Almost all railroads carry both freight and passengers. So how do all these trains travel without crashing?

At first, railroad companies used a timetable system. A timetable shows the time when each train is scheduled to arrive at each station and which trains have the right to move through first.

Today there are more modern ways to keep trains running safely. One way trains stay safe is with the use of traffic signals, including lights. These signals are controlled by people in signaling centers. The signals themselves work a lot like the traffic lights that control cars on the road. They help train operators know about traffic that is coming up ahead of them.

Traffic signals tell trains when to stop and go. They also warn cars when a train is approaching.

compare and contrast

Early train networks only used the timetable system. Why is it more helpful to use signals, too?

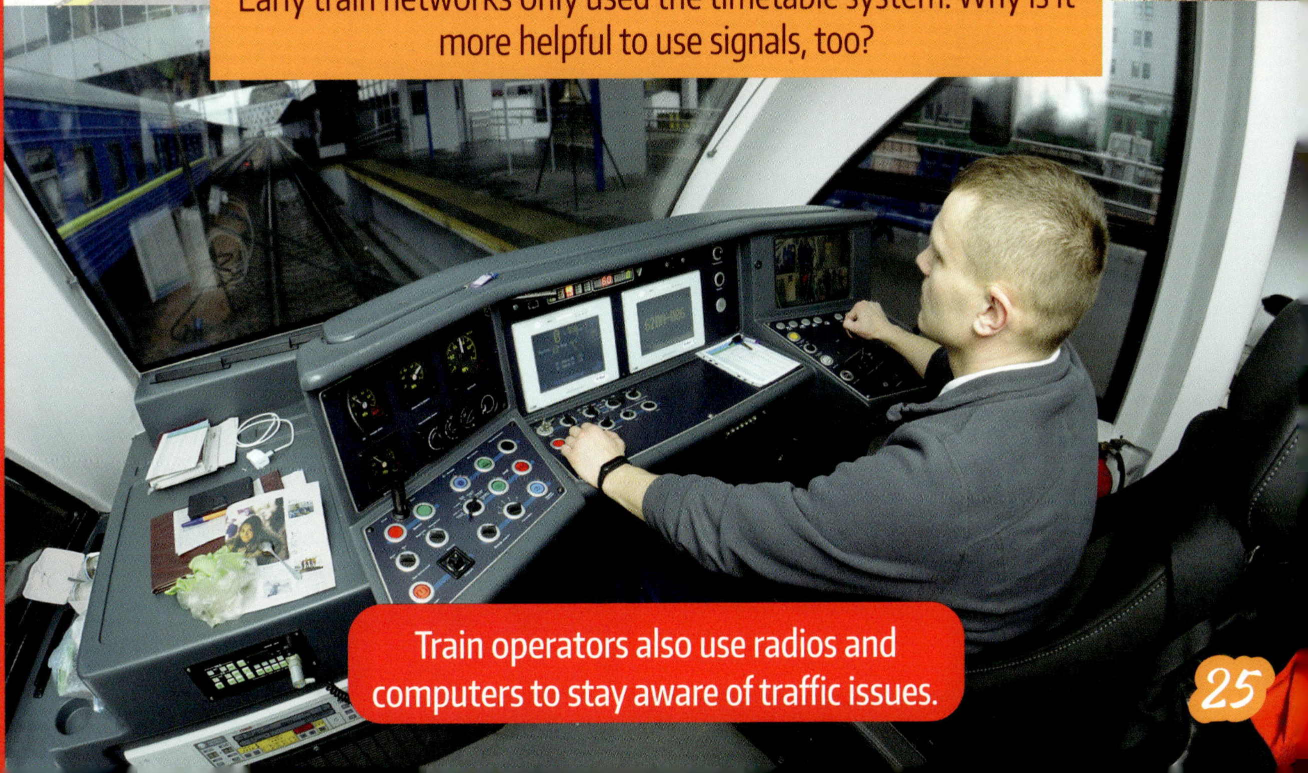

Train operators also use radios and computers to stay aware of traffic issues.

Notable Railways

Some historic railways are famous thanks to their speed, their history, or just the scenes they show off. Here are a few interesting trains.

The Baltimore and Ohio was the first public railroad built in the United States. It began operating in 1830.

Between 1883 and 1977, the Orient Express regularly took passengers from Paris, France, to Istanbul, Turkey. The **luxury** train carried royalty and the very rich.

Since 1898, the Trans-Siberian Express has made the longest regular train trip in the world. It travels from Moscow to Vladivostok, Russia. That's 5,778 miles (9,299 km) in nine days!

You can enjoy your meals in the restaurant car of the Trans-Siberian Express.

The high-climbing Qinghai-Tibet Railway has one of the best views of all train rides in the world.

The Qinghai-Tibet Railway has the world's highest track. It connects Tibet with other parts of China. The train travels over a mountain pass that is 16,640 feet (5,072 m) above sea level.

A train called the Mallard holds the record for the fastest steam locomotive. In 1938 it reached a speed of almost 126 miles (203 km) per hour.

WORD WISE

LUXURY REFERS TO SOMETHING THAT IS DESIRABLE BUT COSTLY OR HARD TO GET.

Take a Trip by Train

Today, people have many choices when it comes to traveling. Many people take airplanes rather than trains for long-distance trips. Others take long ship rides to tropical locations. However, trains continue to be popular ways to travel.

Subways remain one of the best ways to get around cities.

Do you think you'd like to take a ride on a train led by a steam locomotive?

Each year billions of people ride trains to work. In 2013 more than a billion and a half people rode the subways and train lines in New York City, Mexico City, Hong Kong, and Paris. That same year, more than 3 billion people rode the subway in Tokyo and in Beijing!

Other people take trains for fun. In some places it is still possible to ride old steam trains. These are popular for several reasons. Some people just enjoy seeing and feeling what it is like to ride an old train. Others enjoy the sites they can see from the trains.

Consider This

Why do some people still prefer to take a train rather than a plane for a long trip?

Glossary

brake A device for slowing or stopping a moving vehicle such as a bicycle, car, or train.

commuters People who travel back and forth between home and work or school regularly.

freight Goods that are carried from one place to another by vehicle.

gear A device that adjusts the movement of a machine.

levitation The rising or lifting of a person or thing.

locomotive An engine that moves under its own power, especially one that moves cars on a railroad.

parallel Lying or moving in the same direction and staying always the same distance apart.

products Something produced or made.

rails Bars of steel forming a track for wheeled vehicles.

repair To fix or make able to work again.

reverse Go backwards.

route An established, selected, or assigned course of travel.

section A part of a whole.

signals Sounds or motions made to give warning or command.

subways Underground electric railways.

switch A device made usually of movable rails that allows a train or streetcar to move from one track to another.

technology The use of science in solving problems.

throttle A handle or valve controlling the flow of steam or fuel to an engine.

tracks The course along which something moves.

undercarriage A supporting structure or underside.

For More Information

Books

DK Staff. *Train: The Definitive Visual History*. London, UK: Dorling Kindersley, 2023.

Lewis, Norma. *Connecting the Coasts: The Race to Build the Transcontinental Railroad*. Mankato, MN: Capstone, 2024.

Websites

First Transcontinental Railroad

www.ducksters.com/history/westward_expansion/first_transcontinental_railroad.php

Read about the fascinating history of the joining of railroads in Promontory, Utah.

High-Speed Rail Facts for Kids

kids.kiddle.co/High-speed_rail

Read more about high-speed rail travel here.

Index